Race Cars, Muscle Cars, Classic Cars Mosaic Color By Number Adult Coloring Book

By Color Questopia

Copyright © 2020

All rights reserved. No part of this publication may be reproduced, distributed, or transmitted in any form or by any means, including photocopying, recording, or other electronic or mechanical methods, without the prior written permission of the publisher

1. Yellow
2. Black
3. Orange
4. Dark Red
5. Light Orange
6. Dark Brown
7. Brown
8. Dark Gray
9. Red
10. Light Yellow
11. Blue
12. Violet
13. Light Violet
14. Dark Blue
15. Light Green
16. Light Gray
17. Gray
18. Green
19. Light Blue

1. Green
2. Dark Green
3. Violet
4. Light Violet
5. Light Pink
6. Pink
7. Red
8. Light Brown
9. Light Gray
10. Dark Gray
11. Light Yellow
12. Yellow
13. Blue
14. Orange
15. Gray
16. Light Blue
17. Baby Blue

1. Yellow
2. Orange
3. Dark Brown
4. Violet
5. Light Red
6. Light Orange
7. Dark Gray
8. Light Pink
9. Black
10. Dark Violet
11. Dark Blue
12. Blue
13. Gray
14. Red
15. Green
16. Light Gray
17. Light Blue
18. Baby Blue

1. Dark Brown
2. Yellow
3. Light Orange
4. Dark Gray
5. Orange
6. Light Brown
7. Blue
8. Pink
9. Violet
10. Gray
11. Dark Blue
12. Dark Pink
13. Light Violet
14. Light Gray
15. Light Green
16. Light Blue
17. Baby Blue

1. Black
2. Yellow
3. Green
4. Dark Brown
5. Light Brown
6. Dark Blue
7. Dark Gray
8. Red
9. Light Orange
10. Light Red
11. Blue
12. Light Green
13. Dark Green
14. Gray
15. Light Violet
16. Violet
17. Light Gray
18. Light Blue

1. Black
2. Yellow
3. Light Green
4. Light Orange
5. Dark Orange
6. Dark Gray
7. Orange
8. Blue
9. Light Brown
10. Red
11. Gray
12. Pink
13. Light Violet
14. Light Gray
15. Light Blue
16. Dark Pink
17. Violet

1. Light Gray
2. Gray
3. Dark Gray
4. Dark Orange
5. Light Brown
6. Red
7. Light Pink
8. Yellow
9. Light Yellow
10. Blue
11. Light Orange
12. Orange
13. Violet
14. Light Violet
15. Dark Brown
16. Brown
17. Light Blue

1. Black
2. Light Gray
3. Dark Gray
4. Yellow
5. Light Brown
6. Light Red
7. Blue
8. Light Green
9. Green
10. Gray
11. Orange
12. Light Pink
13. Dark Green
14. Dark Blue
15. Red
16. Pink
17. Light Blue

1. Black
2. Dark Brown
3. Yellow
4. Dark Gray
5. Dark Red
6. Blue
7. Dark Orange
8. Red
9. Light Orange
10. Orange
11. Gray
12. Light Gray
13. Green
14. Pink
15. Light Pink
16. Dark Blue
17. Light Blue

1. Brown
2. Dark Orange
3. Dark Gray
4. Black
5. Yellow
6. Blue
7. Orange
8. Dark Blue
9. Red
10. Pink
11. Light Pink
12. Violet
13. Light Violet
14. Green
15. Gray
16. Light Gray
17. Light Blue

1. Light Orange
2. Dark Red
3. Pink
4. Blue
5. Dark Orange
6. Gray
7. Dark Gray
8. Light Yellow
9. Red
10. Dark Brown
11. Brown
12. Light Brown
13. Light Gray
14. Light Violet
15. Violet
16. Yellow
17. Light Blue

1. Orange
2. Dark Blue
3. Dark Gray
4. Black
5. Light Pink
6. Light Yellow
7. Brown
8. Light Orange
9. Red
10. Dark Blue
11. Blue
12. Gray
13. Green
14. Light Violet
15. Violet
16. Light Gray
17. Light Blue

1. Light Yellow
2. Light Gray
3. Dark Brown
4. Dark Gray
5. Black
6. Yellow
7. Brown
8. Pink
9. Dark Blue
10. Orange
11. Blue
12. Dark Orange
13. Gray
14. Green
15. Light Brown
16. Brown
17. Light Blue

1. Light Yellow
2. Dark Brown
3. Yellow
4. Dark Blue
5. Blue
6. Violet
7. Red
8. Light Red
9. Light Gray
10. Light Brown
11. Brown
12. Light Orange
13. Orange
14. Dark Gray
15. Gray
16. Green
17. Light Blue
18. Baby Blue

1. Light Pink
2. Brown
3. Dark Brown
4. Light Gray
5. Dark Gray
6. Black
7. Light Yellow
8. Yellow
9. Red
10. Blue
11. Light Brown
12. Gray
13. Light Orange
14. Orange
15. Light Violet
16. Violet
17. Green
18. Light Blue
19. Baby Blue

1. Brown
2. Black
3. Yellow
4. Dark Orange
5. Light Gray
6. Dark Gray
7. Orange
8. Blue
9. Light Orange
10. Dark Blue
11. Light Pink
12. Pink
13. Light Violet
14. Violet
15. Red
16. Green
17. Light Blue

1. Black
2. Yellow
3. Light Gray
4. Dark Gray
5. Dark Blue
6. Red
7. Light Yellow
8. Blue
9. Violet
10. Dark Violet
11. Gray
12. Dark Green
13. Green
14. Light Green
15. Pink
16. Light Orange
17. Orange
18. Light Blue

1. Yellow
2. Light Violet
3. Brown
4. Black
5. Dark Gray
6. Violet
7. Blue
8. Red
9. Orange
10. Light Orange
11. Light Pink
12. Pink
13. Light Brown
14. Light Gray
15. Dark Green
16. Green
17. Gray
18. Light Blue

1. Black
2. Gray
3. Dark Gray
4. Yellow
5. Dark Red
6. Red
7. Blue
8. Orange
9. Light Pink
10. Pink
11. Light Violet
12. Violet
13. Light Brown
14. Brown
15. Light Gray
16. Green
17. Light Blue

1. Black
2. Brown
3. Dark Blue
4. Red
5. Dark Red
6. Dark Gray
7. Dark Orange
8. Orange
9. Blue
10. Yellow
11. Green
12. Light Green
13. Dark Green
14. Pink
15. Light Gray
16. Gray
17. Light Blue

ENJOY BONUS IMAGES FROM SOME OF OUR OTHER FUN COLOR BY NUMBER BOOKS!

FIND ALL OF OUR BOOKS ON AMAZON

Beautiful Cities and Landmarks
Color by Number
Mosaic World Geography
Coloring Book For Adults

1. Brown
2. Dark Brown
3. Medium Brown
4. Dark Orange
5. Orange
6. Yellow
7. Light Orange
8. Medium Orange
9. Light Browen
10. Light Gray
11. Dark Gray
12. Medium Gray
13. Gray
14. Blue
15. Sky Blue
16. Light Pink
17. Light Violet

Dragon Fantasy
Mosaic Color By Number
Mythical Magic and Lore for Stress Relief

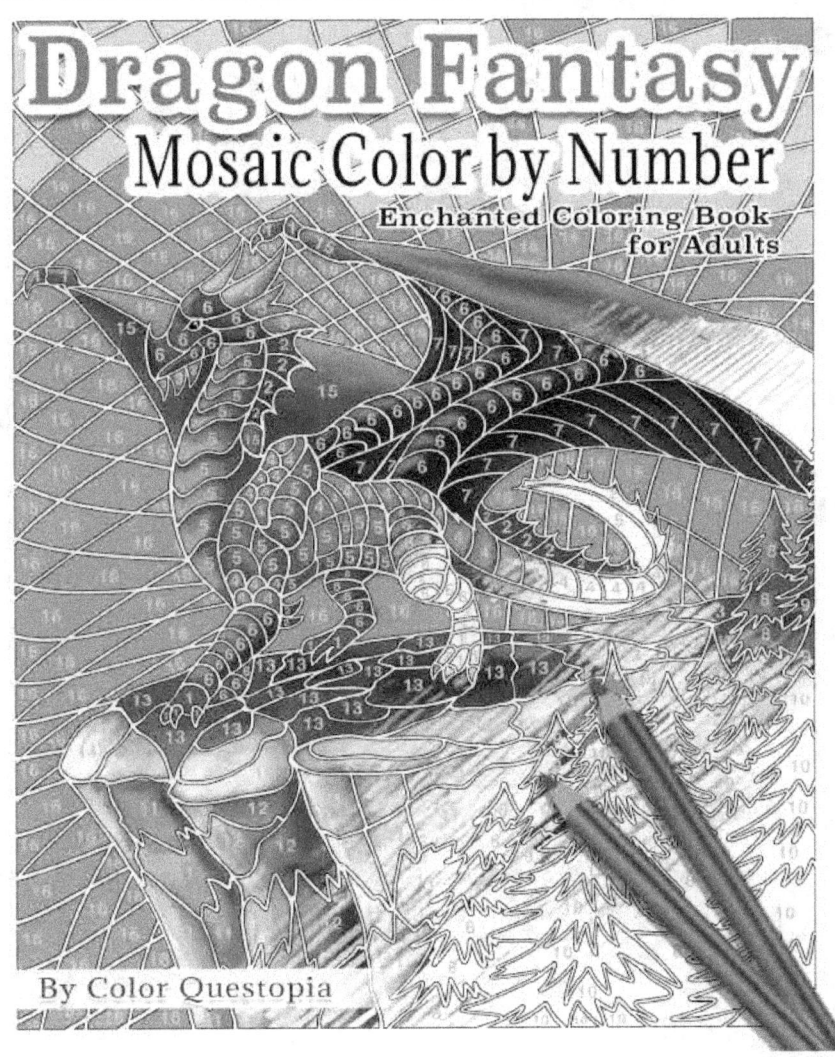

1. Dark Brown
2. Light Pink
3. Red
4. Orange
5. Dark Orange
6. Dark Yellow
7. Light Orange
8. Yellow
9. Brown
10. Army Green
11. Light Brown
12. Blue
13. Light Green
14. Violet
15. Sky Blue
16. Baby Blue
17. Light Blue

Fanciful Fox
Mosaic Adult Color by Number Book
Adult Coloring Book for Stress Relief
and Relaxation

1. Black
2. Dark Brown
3. Light Orange
4. Light Yellow
5. Light Brown
6. Orange
7. Beige
8. Dark Yellow
9. Medium Brown
10. Brown
11. Gray
12. Dark Gray
13. Light Green
14. Green
15. Medium Gray
16. Light Gray
17. Sky Blue

New York
Mosaic Color by Number
Coloring Book for Adults

1. Yellow
2. Light Yellow
3. Brown
4. Light Pink
5. Dark Orange
6. Pink
7. Orange
8. Violet
9. Light Brown
10. Dark Violet
11. Medium Violet
12. Dark Brown
13. Dark Blue
14. Light Blue
15. Blue
16. Sky blue
17. Gray
18. Light Gray

Country Farm Scenes
Nature, Animal, and Easy Designs
Adult Coloring Book
Color By Number For Adults

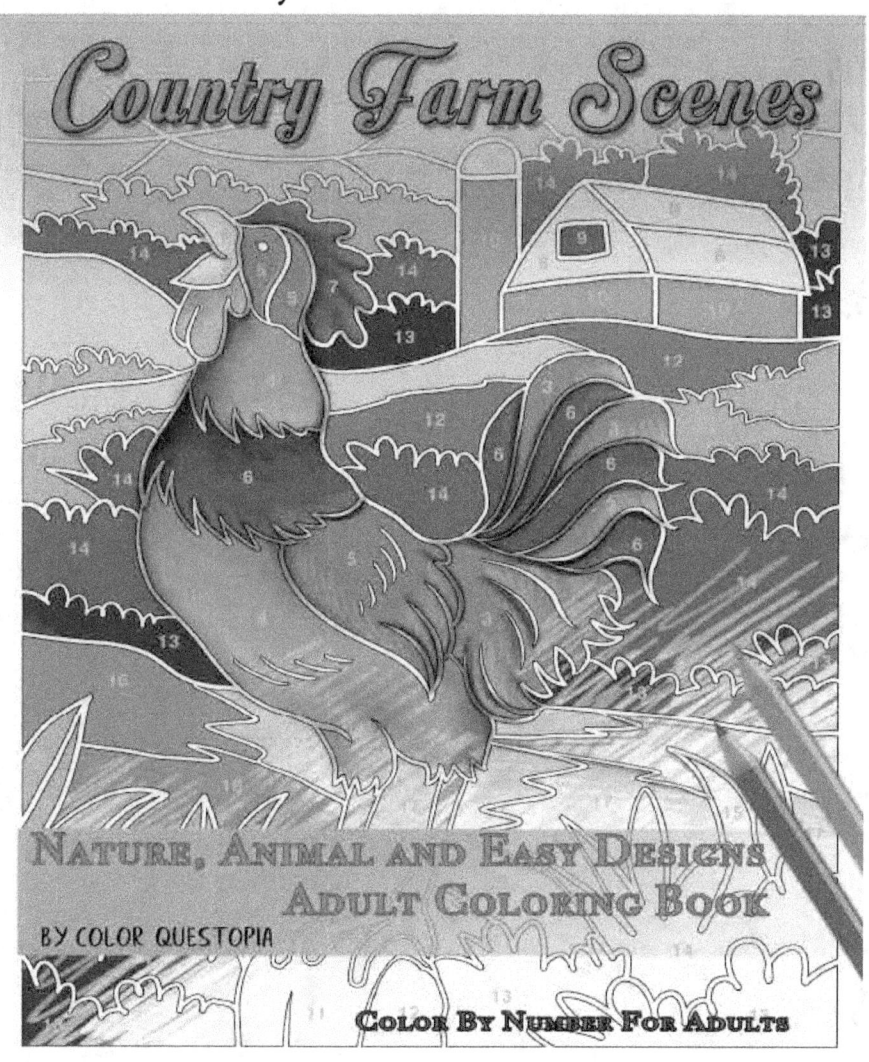

1. Red
2. Dark Red
3. Pink
4. Dark Brown
5. Red
6. Yellow
7. Dark Green
8. Medium Green
9. Army Green
10. Green
11. Light Green
12. Light Brown
13. Brown
14. Gray
15. Light Gray
16. Blue
17. Light Blue

www.ingramcontent.com/pod-product-compliance
Lightning Source LLC
Chambersburg PA
CBHW081500220526
45466CB00008B/2727